Little Book Big Universe

A GUIDE TO EXPLORING AUSTRALIA'S NIGHT SKIES

Professor
LISA HARVEY-SMITH
Illustrated by First Dog on the Moon

To Mum and Dad,
who encouraged my
eccentric hobbies.

First published in Australia in 2021
by Thames & Hudson Australia Pty Ltd
11 Central Boulevard, Portside Business Park
Port Melbourne, Victoria 3207
ABN: 72 004 751 964

thamesandhudson.com.au

24 23 22 21 5 4 3 2 1

Thames & Hudson Australia wishes to acknowledge that Aboriginal and Torres Strait Islander people are the first storytellers of this nation and the traditional custodians of the land on which we live and work. We acknowledge their continuing culture and pay respect to Elders past, present and future.

ISBN 978-1-760-76229-2

A catalogue record for this book is available from the National Library of Australia

Illustration: First Dog on the Moon
Design: Philip Campbell Design
Editing: Paul Smitz
Printed and bound in Australia by McPherson's Printing Group

FSC® is dedicated to the promotion of responsible forest management worldwide. This book is made of material from FSC®-certified forests and other controlled sources.

Contents

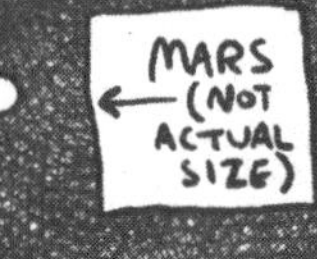
MARS
(NOT
ACTUAL
SIZE)

1
A universe bursting with stars

Hello, fellow Earthlings. My name's Lisa and I'm a scientist. An astrophysicist, to be precise. My job is to study really cool stuff like stars, galaxies and black holes using HUUUUUUGE telescopes. As part of my scientific adventures, I have been lucky enough to discover some incredible secrets about outer space.

Wanna hear one? Okay, try this on for size:

Right now, somewhere in the cavernous depths of space, a star is bursting into life.

A star is born

It might be hard to imagine, but every day stars are born and they die, just like us. Except that the birth of a star is quite different to the way you and I started our lives. We began as scrunched-up, wrinkly babies kicking and screaming our way into the world (what a racket!). But we expect none of that sort of fuss and nonsense from our twinkly buddies in the sky.

Unlike a bawling infant, a baby star peacefully takes shape inside a gigantic cloud of gas. First, the gas is moulded into a ball by the force of gravity. The ball gets smaller and smaller until the middle of the star heats up to many millions of degrees Celsius. The hydrogen gas in its belly turns into helium, in what we call nuclear fusion, and suddenly, SCHWOOOM! – the star erupts in a shower of heat and light, beginning its long stint as a shining beacon in the sky. It's a supremely powerful event, and it happens in complete silence.

This incredible process of star birth is repeated millions of times every day across the depths of space. We estimate that across the universe (that's the word for everything that exists), around 10,000 new stars burst into action every second.[🚀] Just think about that. Pop! Pop! Pop! Pop! Imagine how beautiful it would be to see thousands of new stars exploding into light above your head in the blink of an eye.

So why *don't* we see thousands of new stars appearing left, right and centre? It is simply because of the size of the universe. Space is so humungous that the light from faraway stars is too faint to see.

Two hundred billion stars make up our galaxy, the Milky Way. It's an unimaginable number really. To count them all, it would take you at least 3000 years.

🚀 Based on the number of new stars that appear in our own galaxy each year (around 3) and the number of galaxies in the observable universe (around a hundred thousand million).

'One hundred and ninety-nine thousand million, nine hundred and … GAH! I've lost count!!

Sigh

One, two, three …'

Beyond the Milky Way there are probably more than 1 million million million million (1,000,000,000,000,000,000,000,000) stars. That's not to mention the even bigger numbers of planets, asteroids, comets, moons, meteors, gas clouds, galaxies and black holes that populate the vast expanses of space.

We can't see all the stars in the Milky Way, let alone those that lie trillions of kilometres away, far beyond our cosmic horizon. These regions only come into focus when we look up to the sky with huge telescopes.

But there are many fantastic things you can see inside the Milky Way with just your eyes and a bit of patience. If you add a pair of binoculars or a small telescope, and a dark sky, you will open up a virtual portal to travel through space and time.

See for yourself

I started exploring the universe one night when I was 12 years old, in my backyard with my dad. It was very cold and I pretended not to be scared of the dark (and what I imagined was lurking in the shadows). Armed only with an old-fashioned map of the stars, Dad and I figured out which way was north and south. Suddenly, what we were seeing in the sky made sense, and before long we were discovering constellations, star clusters, meteors and other remarkable sights.

That night we saw the planet Mars shining among the stars like a bright orange button.

Later that year, I sat in a deckchair in the dark as a meteor shower rained shooting stars down the inky blackness of the evening sky. I soon figured out how to find the Andromeda galaxy, a cosmic city containing a trillion glittering jewels whose light has travelled for 2 million years to reach Earth. Space was so full of possibilities. Once I unlocked its secrets, I never looked back.

Now I want to share this journey of exploration with you. Read on and learn how to hunt for planets, photograph star trails and look back in time. You too will become expert at viewing meteor showers, exploring the craters on the Moon, and waving to the astronauts aboard the International Space Station! Those (like me) who enjoy some extra information can read the notes at the bottom of the page to learn more.

Australia's beautiful night skies are yours to discover. All you have to do is look up.

MOZ-E GO!
COOL BOX

2

Come to the dark side (but don't forget mozzie spray)

Before we launch into outer space, let's start with some essential mission training. Here are some basic tips on making the most of your astronomical adventures.

The first thing sounds obvious, but the best time for stargazing is ... wait for it ... night-time.

Woah. Hold it right there, Einstein!

Have you ever thought about how the stars lay over Earth in every direction like a beautiful sparkly cloak? The stars are not just above our heads but also below our feet, on the other side of the world. They are always in the sky – even during the day. But we can't see the stars in the daytime because our sky is filled with sunlight, which bounces off the tiny particles called molecules that make up our atmosphere. Our eyes are so completely overwhelmed by the brightness of the sky that fainter objects like the stars become invisible to the naked eye.

Seeing in the dark

Our eyes are astonishing, adaptable machines. In low light conditions, they slowly get used to the darkness, giving you super-powerful 'night vision'. This happens because your pupils (the dark circles in the middle of your eyes) expand to let in more light. A chemical called rhodopsin, which is released in the light-sensitive part of your eye (called the retina), turbocharges your ability to see in the dark. Spending at least 20 minutes in a dark place before you head outside at night will allow your eyes to adjust and become up to a million times more sensitive! That's a whoooooole lot of extra stars, meteors and satellites you're going to see.

Light pollution is another thing to avoid when you're finding your way around the night sky. Towns and cities are awash with 'sky glow' – light that's scattered into the sky by badly designed streetlights, floodlights and even car headlights. Light pollution wastes energy and money and contributes to climate change. It is also damaging to creatures that rely on Earth's natural cycles of night and day to feed, breed and migrate.🚀

A dark sky without light pollution is a breathtaking sight. It is much better looking at the stars from a dark

---

🚀 To join the campaign against light pollution in Australia, visit australasiandarkskyalliance.org.

place like the countryside, or better still the outback. If (like most Australians) you live in a town or a city, ask an adult if you can head somewhere darker like an oval, a national park or the coast. When I go stargazing, I usually find a dark beach or a cleared area of bushland. It makes a huge difference to what I can see compared with when I'm standing in my backyard.

Check out an online dark-places map to pick an ideal stargazing location in your area – try the one at darksitefinder.com.

In the right place at the right time

Look after your safety when you're out stargazing. Always take a grown-up with you and be careful not to trip on hidden rocks, tree roots or other hazards. A piece of flat, open ground away from buildings or trees is best, so you can see the whole sky. Remember that your night vision superpowers will quickly disappear if you look at any bright lights, so stay away from nearby roads, since car headlights can dazzle your eyes. If you need to light the way to your chosen stargazing spot, take a red light or a torch with a piece of red cellophane stuck across the beam – your eyes are less sensitive to red light, so it won't ruin your night vision.

Darkness is much easier to find during the winter months because the nights are longer. In June, July and August, Australia is pointed slightly away from the Sun, and in most parts of the country it gets dark by dinnertime. By mid-evening, stars decorate the sky like the twinkling lights of Christmas.

In summer, Earth's Southern Hemisphere (that's us) is pointed towards the Sun and the days are longer – in December it is probably still light when you go to bed. If you want to stargaze during the summer months, you'll have to stay up later (after 9 pm in northern Australia or 10 pm in the south of the country) or wake up very early in the morning (before 5 am in the north and 4 am in the southern states). Although the nights are shorter in summer, there is still plenty to see. It's my favourite time to watch the full Moon rising over the ocean, or satellites soaring overhead (more about this in Chapter 7).

Be aware of the phases of the Moon – the full Moon is 32,500 times brighter than the brightest star you can see. If you want to enjoy the stars in all their glory, you definitely want to avoid the 2 or 3 days either side of a full Moon. You can check out the current lunar phase at timeanddate.com/moon/phases.

Finally, stargazing requires a bit of patience, so look after your comfort when you're out on a star-finding mission. Camping chairs are a great idea. I also often take a hot drink and some snacks when I'm star-watching.

Make sure you have plenty of warm clothes, depending on the conditions, and don't forget mozzie protection – you'll enjoy the experience so much more if you don't come back covered in annoying itchy red bumps and spend the next day scratching yourself like a kangaroo!

3
Star maps, apps and other fun gadgets

Getting to know the night sky is like exploring a new neighbourhood. You've got to spend some time getting your bearings, and if you get stuck, reach for a map. If you have access to a smartphone or tablet, I highly recommend that you download a night sky or stargazing app. I use Sky Guide, which costs $4.49 and is well worth the money, but SkyView Lite is free.

A good stargazing app will show you the names of all of the brightest stars in the sky, and the positions of planets, satellites, galaxies, star clusters and constellations. Constellations are groups of stars that our ancestors named after mythical animals, people or objects. Even today, constellations are a useful way of navigating through the night sky and finding interesting things. I call it star hopping – once you've learned to recognise one bright star or constellation, you can more easily follow the shapes in the sky to the next one. It opens up the whole universe for you to explore.

A night sky app works by using the GPS location services in your device and in-built sensors to calculate where you are and in which direction you are pointing. With a couple of clicks, a dynamic starscape appears on-screen, moving and changing as you hold the device in front of you. And if you take it on a camping trip in your area, or head interstate or even overseas, the location and time zone automatically update and you can explore the sky wherever you might be. It's great!

If you don't have access to a smartphone or tablet, you can find a map of the night sky that's updated monthly at the Sydney Observatory or Perth's Scitech museum websites – pick the one closest to you for an accurate guide to the sky where you are. Or you can get yourself a 'planisphere': a small, handheld plastic disc showing the stars that are visible at different times of the year. To get an accurate picture of the stars on a particular night, rotate the little dials on the outside of the disc until the arrows point to that day's date. Just make sure that if you are buying one online, it's a Southern Hemisphere planisphere. If you buy a Northern Hemisphere version, it will show you the stars overhead in Europe, North America and Asia, many of which are invisible from Australia.

Smile for the camera!

Astrophotography (photographing the stars) is a really fun hobby and leaves you with beautiful images to show your friends and family. You will get great results with a digital SLR[*] camera (that's a big, bulky camera like a professional photographer would use), but I just use my smartphone and an app called NightCap Camera. I find this easy to use and ideal for capturing images of stars, planets, star trails and even the aurora australis (the southern lights). For the best results, you'll want to use a tripod to hold the camera still – you can buy a smartphone tripod online for about $20. If you don't have a tripod, then prop the camera up against a strong, immovable surface so that you can point the device up at the stars.

If you have a digital SLR camera, that's tremendous! Set the camera to the manual setting (select 'M' on top of the camera) so you can leave the shutter open for as long as you like. Turn the flash off, because we don't want to create our own light pollution and ruin the picture. Turn autofocus off and manually focus to infinity by twisting the lens fully until the stars appear completely sharp. Set the aperture (the width of the opening that lets light into the lens) to the lowest f number (for example, f/2.8), which opens the camera's 'eye' wide to let as much light in as possible. Now set the ISO (sensitivity) to a high setting like 800 or 1600.

[*] SLR = single lens reflex.

You can experiment with this, since the best setting depends on the camera. If you set it too high, your pictures will look grainy. Finally, point the camera up at the stars and press the shutter button to take a picture. Now wait.

On the manual setting, the shutter will stay open for as long as you want. Press the shutter button again and your picture is done. Try exposure times of between 30 seconds and 60 seconds to capture the stars, planets and Milky Way. A longer exposure of 10 minutes, 20 minutes or even an hour will capture spectacular star trails as Earth rotates. Don't be afraid to experiment with a range of different settings.

When you are using a night sky photography app on a smartphone, the rules are much the same. If you have NightCap Camera, open the app and touch the star shape in the bottom left-hand corner of the screen. A number of different camera options will pop up. I like to select the 'long exposure' mode so I can fully control the camera's settings.

Then select the settings icon (top of screen, looks like a wheel). Slide the Interval Programmer on and set the exposure (EXP) to 30 seconds. This means that when you press the camera shutter button, it will collect light for 30 seconds. Press the settings icon again to close this menu, then place your smartphone on a tripod or a flat surface, aim it at the stars and take your picture. You may need to play with the settings and tap the screen at the point you want to focus (the stars). Give it a go – it's great fun. You might also want to try the automatic stars and star trails settings, or the ISS (International Space Station) or meteor modes to capture satellites or shooting stars (read chapters 7 and 9 for more details).

Scoping out the right tools

You don't need a telescope to enjoy looking at the stars, but if you are lucky enough to get one it can be a fantastic addition to your stargazing kit. So what kind of telescope do you need? And what can you see?

It's important to get to know the stars and constellations really well *before* you buy a telescope – I'll tell you how to do that in Chapter 4. Unless you know what you're looking for, you'll just point your telescope at the sky and see a close-up view of some random stars, which look like, well ... stars. Disappointment guaranteed. But once you get to know your way around the night sky, the right telescope will open up awesome sights like the rings of Saturn, the moons of Jupiter, and some breathtaking star clusters and nebulae (space clouds).

A common mistake is to buy a really small telescope. They are just not worth the money (I'll say more about this in Chapter 10). The greater the aperture, the more light is collected by the telescope, and therefore fainter and farther-away objects can be seen. If the aperture of the telescope is less than about 7 cm, you're better off with a good pair of binoculars. You can use them to zoom in on star clusters, planets, galaxies and more. A compact pair of birdwatching binoculars is fine for scanning around, but a bulkier pair with a larger aperture will give you great magnification of faint night sky objects. Just make sure you have a tripod. I have a pair of 10 × 50 mm binoculars, which means they make things look 10 times bigger and the aperture of the lenses is 50 mm. You can comfortably see 4 of Jupiter's moons with them, which is really cool. Bigger is generally better with binoculars, so you might want to upgrade to the larger sizes (for example, 25 × 70 mm or 20 × 80 mm) if you can afford to.

That's just about everything you need to head outside and begin your stargazing adventures. Next up, you're going to learn how to recognise the brightest stars and constellations (groups of stars). Let's go!

THE
BONGO

TINKLE
TINKLE
DING

GRANDMA'S
SOCKS

CAT PAWS

HELLO
MONKEY

HONK

PICKLE
JUICE

MOO

NOODLEBOING

4
Constellations and star hopping

We're going to start our star-hopping adventure with the Southern Cross, probably the most recognised constellation in the southern sky. It is known by many names across the world, including Crux (Latin for 'cross'), the Women in the Sky (south-western Western Australia), Mirrabooka (Stradbroke Island), the Stingray (Arnhem Land), the Eagle's Footprint (central Australia), the Anchor (New Zealand) and the Giraffe (southern Africa). The Southern Cross appears on the flags of New Zealand, Australia, Brazil, Papua New Guinea and Samoa.

In the southern parts of Australia, you can find the Southern Cross at any time of the year, but it disappears below the horizon on late spring evenings in the north of the country. To find it, turn your body to face approximately south. Uh oh, where's that? Never fear. Use a compass, a compass app or your night sky app to find the right direction (I am also about to teach you how to do that anywhere, anytime).

The Milky Way, Emu in the Sky and Southern Cross

Now that you are facing south, scan up and down the sky until you find the Milky Way. From a city, the Milky Way looks like a sort of highway above you made up of hundreds of stars. If you are somewhere quite dark, it looks like an amazing band of cloudy white light stretching right across the sky. It also features dark stretches of dust that absorb background starlight, making up the famous dark constellation known as the Emu in the Sky. Found it? Awesome!

The band of stars visible as the Milky Way is our view from inside the flat disc of our home galaxy, which is made up of hundreds of billions of stars, including the Sun, as well as vast quantities of gas and dust. Nestled in the Milky Way in an approximately southerly direction is the Southern Cross, a small constellation tipped by 4 bright stars. It may be smaller than you expect: if you hold your arms out straight and put your thumbs together side by side, you can completely cover the cross.

Now an important note: when you are hunting for constellations, remember that they might appear upright, sideways or upside down, depending on the time of night. That's because Earth is constantly spinning on its axis (the imaginary line joining our planet's north and south poles). This rotation causes our day and night. It also means the stars appear to circle around a fixed point in our sky called the South Celestial Pole, which lies directly above the South Pole. A night sky app will correctly rotate them for you, but if you're using a star map, then don't be surprised if you have to turn your head (or the map) onto its side to see everything.

With this in mind, tilt your head so the cross appears upright like a crucifix. The star at the top is slightly orange in colour. To the left-hand side of the cross (imagining it upright), you'll see 2 very bright stars called the Pointers. If you're ever lost in the bush or at sea, you can use the constellation to find due south. Close one eye and pinch the 'top' and 'bottom' of the cross between your thumb and forefinger. That's your measuring stick. Now measure 4 more of these lengths in a straight line below the cross and ta-da! You're looking due south. That little trick might just save your life one day.

Exploring more southern constellations

From May to November, follow the line from Crux through the Pointers along the Milky Way and you'll eventually come across the distinctive curved shape of Scorpio (the Scorpion), also known as the Fishhook or the Dragon. Scorpio is highlighted by the orange star Antares (pronounced an-*tar*-ees), which means 'rival of Mars'. When you see its vivid colour, you will understand why!

Scorpio borders on the constellation of Sagittarius, the Archer. Some people see part of Sagittarius as the shape of a teapot. Whatever floats your boat. Between the 2 constellations, in the centre of the Milky Way, lurks a supermassive black hole called Sagittarius A* (A-star). You can't see it because black holes are, well, black, but you can stare at the spot and imagine its epic gravitational pull, which is equivalent to 4 million Suns.

From the heart of the Milky Way in Sagittarius, you can practise star hopping to the other bright winter stars. Use your night sky app or planisphere to help you find the orange star Arcturus (arc-*ture*-us) in the constellation of Boötes (*boo*-tease, like the cute footwear for babies). Also intense enough to spot from all stargazing sites are the white stars Spica (*spy*-ka), in the constellation of Virgo, and Regulus (*regg*-you-luss), the brightest star in the constellation of Leo, the Lion. Both of these are actually double stars, with components that are too close to separate with the naked eye.

On the other side of the sky from Scorpio lies the impressive Orion, the Hunter – also known as Baiame, or Sky Father. Orion features prominently in the Australian sky between December and March. Much larger than the Southern Cross, Orion is outlined by a huge rectangle of supergiant stars, each one many hundreds or thousands of times brighter than the Sun (but a lot further away from us). At the bottom right-hand corner of this rectangle is an old, dying red supergiant star called Betelgeuse (*beetle*-juice). Sweeping diagonally through the centre of Orion is a triplet of stars called Orion's Belt.

Keeping Betelgeuse in the bottom right, follow the belt up and to the right and you will arrive at a piercing white star called Sirius – the brightest star in the whole sky. It is often called the Dog Star because it is the most prominent star in the constellation Canis Major (the Big Dog). In the other direction, tracing a line down from Orion's Belt takes you to Aldebaran (al-*deb*-er-an), the eye of Taurus, the Bull. This red giant star is 400 times more luminous than the Sun (but 4 million times further away) and it slowly varies in brilliance over the years.

With so much to see in and around the southern sky, grab a night sky app, a star map or a planisphere and start investigating it for yourself. Take some time to explore before we delve deeper into space to discover the spectacular stars, planets, comets, meteors, satellites and other extraordinary things that soar above us.

5
Planets, rings and moons

Planet spotting is one of my favourite things! There is hardly a night in the year when you can't see at least 1 of our solar system's 7 planets (besides Earth) in the sky.

Planets don't shine with their own light like stars do. They act like mirrors, reflecting the light from the Sun. The sunlight bounces off their rocky surfaces (Mercury and Mars) or cloud tops (Venus, Jupiter, Saturn, Uranus and Neptune) and into our eyes. Despite generating no light of their own, planets are SO much closer to Earth than the stars[*] that many of them outshine even the brightest stars in our sky.

So how can you know which 'stars' in our night sky are actually planets? The best way is to jump online and check out a monthly sky guide for your location, or explore timeanddate.com/astronomy/night to find the precise position of planets in your part of the country tonight. Read on for some tips on how to recognise

[*] Our nearest star (Alpha Centauri) is 13,000 times further away from Earth than Uranus.

them when you're in the field. If all else fails, you can type the name of any star, planet or other celestial object into your star app's search feature and it will guide your exploration.

Let's take a virtual tour through our solar system to find out what you can expect to see.

A terrific trio

The smallest planet is Mercury, a hot rocky world that speeds around the Sun once every 88 days. Because it hugs the Sun so closely (Mercury is 3 times closer to the Sun than we are), we only see the planet shortly before sunrise or just after sunset. It looks like a crisp point of white light peeking out of the yellowish twilight. Since the Sun sets approximately in the west, you will need to choose a stargazing place with a clear horizon in that direction to see Mercury. I usually pick my front garden, or an upstairs bedroom window, with the lights off of course! If you are a morning person, peek out towards the east before sunrise. The best times to catch Mercury over the next year are September 2021, February to April 2022, and June, August, October and December 2022. The planet's position changes quickly, so check your night sky app for the precise location of this elusive world.

The next planet from the Sun is Venus, a most spectacular sight in our skies. Like Mercury, Venus's orbit is relatively close to the Sun, so for much of the year it is lost in our star's overwhelming glare. But when it appears at maximum separation from the Sun, Venus dominates our eastern skies before sunrise, or it gleams above the western horizon after sunset. At its most spectacular, Venus appears 17 times brighter than Sirius – it's dazzling, like a brilliant jewel. The best times to spot Venus will be in the evening twilight (after sunset, as it starts to get dark) between September and December 2021, then in the morning twilight from February to July 2022.

If you have a fairly good telescope, you will be able to see the 'phases' of Mercury and Venus change as the planets orbit the Sun. Just like the Moon, the area illuminated by the Sun changes from a thin crescent to half the face and eventually the full face of each planet as they move in their orbits.

Mars is the fourth planet from the Sun (after Earth). Known as the Red Planet, its striking colour is thanks to the large quantities of iron oxide (rust) contained in its dirt, similar to many outback regions of Australia. Mars has only a very thin atmosphere and little cloud cover, meaning that sunlight is reflected from the surface and the planet appears orangey-red. The planet will be well placed in Australian skies from January to December 2022 for naked-eye observation. If you have access to a telescope, you will sometimes see the whites of ice caps, formed when carbon dioxide from Mars's thin atmosphere freezes onto the planet's polar regions during winter.

Planets that are far out!

Jupiter is next in our space odyssey, and this giant planet makes a great target for budding cosmic explorers. Made mostly of gas and liquid, Jupiter has a small rocky core buried at its centre. Sunlight reflects off the swirling clouds in the planet's upper layers to create a brilliant star-like white globe of light for the unaided eye. Through a good pair of binoculars held steady on a tripod, it is possible to see up to 4 of Jupiter's largest moons – Io, Europa, Ganymede and Callisto – which appear as tiny star-like points of light beside the planet. With a large telescope, you can see the beautiful coloured bands of Jupiter's clouds. Get down to your local public observatory or astronomy club to enjoy this wonderful sight in Australia's skies between September 2021 and January 2022, and again from April to December 2022.

Now we come to the second giant planet, Saturn. Like Jupiter, Saturn is made largely of gas. Even at a distance of 1.5 billion km from the Sun, Saturn appears as a bright yellowish star-like object to the naked eye. Binoculars won't be help much for studying the planet in detail, but a small telescope with a 100 mm aperture will allow you to see the famous rings of Saturn, made up of billions of boulders, rocks, dust grains and pieces of ice that circle the planet. A larger telescope (200 mm aperture or greater) will show you spectacular detail of the intricate ring structure and the dark opening in Saturn's rings called the Cassini Division.

The planet has more than 60 moons, the largest of which, Titan, can be spotted using a good pair of 10 × 50 mm binoculars. Through a modest-size telescope (150 mm aperture), you should be able to spot 7 of Saturn's largest moons. The brightest of these orbit the planet in a matter of hours or days – to identify which moons you are seeing, manually zoom in on the planet on the Sky Guide app. Once you have scanned in close enough, the names and positions of the moons will appear on the map. Saturn will be well placed for observation in the late evening skies from September to December 2021. It will then be most visible through the early hours of the morning from March to May 2022, before returning to evening skies across the country from June to December 2022.

The outermost planets in our solar system are Uranus and Neptune. These cold worlds orbit in the depths of space, far from the warming influence of the Sun. Both can be spotted through binoculars, if you know where to look (hello, app!). Uranus will be visible in the southern skies from September 2021 until January 2022, then again from August to December 2022. It looks like a blueish disc of light.

Neptune is fairly difficult to spot due to its huge distance from the Sun (around 4.5 billion km, give or take a few!). It is several times fainter than Uranus, but that doesn't mean you shouldn't try to see it. The best times to attempt a glimpse are in September 2021 and September 2022, when the planet will be directly opposite the Sun and visible through binoculars or a telescope. It will appear a darker blue than Uranus, and will look like a tiny dot due to its vast distance from Earth.

6
The great cratered sky-crab

The Moon is the ultimate shape-shifter. On its 28-day journey around Earth, our planet's natural satellite presents alternating faces to the Sun. At the full Moon, when completely illuminated by the Sun, it washes our night sky with its eerie cold light, blotting out all but the brightest stars and planets. A week later we see a half-Moon, a semi-circular slice of the Moon that's lit up during the daytime. Over the next few days the Moon retreats into a slender crescent and finally disappears into the Sun's glare as a 'new Moon' – a completely dark body. Then the cycle begins again, with the thin crescent Moon getting fatter as it advances into the evening sky.

Lunar exploration

Look carefully and you will notice a collection of dark patches on the surface of the Moon. People in the Torres Strait describe these shapes as a local woman weaving

A satellite is an object that orbits another object. The Moon is Earth's only natural satellite, the others being orbiting spacecraft made and launched by humans.

a mat. European and Asian mythologies talked about the woman or man in the Moon – someone who was supposedly banished to the Moon after breaking the law. More recently, Europeans named the dark shapes 'seas' (or maria in Latin), although we now know that the Moon is a very dry place. Personally, I think they look like a crab. What can you see in the Moon?

Today, we understand far more about the Moon because of the many spacecraft that have been sent to study it. It turns out that the dark patches are ancient craters caused by collisions with space rocks and which then filled up with volcanic lava, more than a billion years ago. The rocks contain iron, which gives the maria their dark colour. In contrast, the other parts of the Moon's surface (called the highlands) are much brighter.

The lunar maria have some pretty cool names. You may have heard of the Sea of Tranquility, where the first people to walk on the Moon – Neil Armstrong and Buzz Aldrin – landed aboard Apollo 11 in July 1969. Other lunar features include Oceanus Procellarum (the Ocean of Storms), Mare Crisium (the Sea of Crises) and Mare Nubium (the Sea of Clouds). These rather imaginative descriptions conjure up an image of a cloudy, stormy Moon, when in fact it is dry, cloudless and airless!

'Maria' is pronounced '*marr*-ee-uh'. The singular form is 'mare' (*mah*-ray).

You can explore the lunar 'seas' and the Moon's other features on a map made by the Lunar Reconnaissance Orbiter Camera, an orbiting spacecraft that captures close-up pictures. Go to quickmap.lroc.asu.edu and select the 'nomenclature' overlay to see the names of maria and craters. The default map view shows the Moon as it appears in the Northern Hemisphere – pick the Lunar Globe (3D) projection so you can click and drag the image to line up the Moon as it appears in the Australian sky.

Look at the top of the full Moon and you'll see the Tycho (*tie*-ko) impact crater, which was caused by a large space rock (meteorite) that collided with the Moon some 108 million years ago. Using binoculars or a telescope, you'll see the lunar surface covered in thousands of craters – the result of billions of years of collisions. With no wind or running water to erode the surface, the Moon's craters have remained, a timeless reminder of an eventful past. There are also hundreds of mountains and valleys to explore. Choose a night when there's a half-Moon or our satellite is crescent shaped, so that the long shadows highlight these features.

Photographing the Moon is fun and challenging. Set up a pair of binoculars or a telescope and point your smartphone camera down the eyepiece. So long as you hold the phone steady (you can get an adaptor) and your camera is able to focus on the Moon (tap the screen on the Moon to focus), you should get some pretty good shots. If the Moon appears over-exposed (i.e. a big bright blob), reduce the ISO and the exposure time on the NightCap Camera app until you hit the sweet spot. If you don't have a telescope or camera, you can draw the Moon's dark patches and identify the features using a lunar map.

Epic eclipses

Every now and again, the Moon lines up perfectly with Earth and the Sun to create an eclipse.[*] Eclipses happen because of an extraordinary coincidence: the Moon is about 400 times smaller than the Sun, yet it is also 400 times closer to Earth, so the Moon and the Sun appear almost exactly the same size in the sky.

A total lunar eclipse happens when the Moon enters the darkest part of Earth's shadow. When this happens (on average every 2.8 years in Australia), the full Moon turns blood-red due to the light scattered through Earth's atmosphere. At its darkest point, it becomes eerily black. A partial lunar eclipse occurs when the Moon is only partly obscured by our planet's shadow.

---

[*] This would happen twice a month if the orbits of the Moon and Earth were perfectly circular. But they are not, so the 3 bodies – Earth, Moon and Sun – only line up perfectly every so often.

During a lunar eclipse, you can watch Earth's shadow creep across the face of the full Moon in real time. The colours can be amazing too! The next opportunities for seeing one are the partial lunar eclipse of 19 November 2021 and the total lunar eclipse of 8 November 2022, both of which are visible from Australia's shores. Try capturing a series of photos of the Moon before, during and after the eclipse. It is a fantastic sight.

Solar eclipses (where the Sun disappears behind the Moon) are far rarer because they only happen across a very thin slice of Earth's surface. That means a solar eclipse will rarely come to you.

Many people choose to travel to see solar eclipses and often make a holiday of the event. When I was 19 years old, I went with some friends on a camping trip to Cornwall, a pretty coastal region of south-western England. It was 1999, and we wanted to witness the total solar eclipse that was going to darken skies across a thin stretch of western Europe, the Middle East and India on 11 August.

On that day, we awoke to dense cloud cover. Tiny breaks emerged throughout the morning and we managed to get glimpses of the crescent Sun using our special eclipse glasses.🚀 Finally, the epic moment

---

🚀 Never, ever look directly at the Sun, especially through binoculars or a telescope. You may cause permanent damage to your eyesight.

of totality arrived, when the Moon completely covered the Sun. Alas, thick clouds had returned and we missed seeing the spectacular 'solar corona', or super-hot atmosphere of the Sun. But in those 2 minutes of darkness, the wind dropped, the air cooled, and hundreds of enthusiastic sky-watching campers gasped and wow-ed. I will never forget it.

Australia will be treated to a couple of total solar eclipses in the coming years. During the next one on 20 April 2023, a tiny coastal region of Western Australia near Exmouth will experience an incredible show. For those lucky enough to witness it, the Moon will slowly cover and uncover the Sun, with the whole process lasting around 3 hours from start to finish. The moment of totality will only last about 60 seconds, but it will be a breathtaking sight.

The next total solar eclipse that will be visible across large parts of Australia will occur on 22 July 2028. It should be a magnificent event for millions of people, as its path crosses greater Sydney, parts of Queensland, the Northern Territory and Western Australia. How old will you be when the next great Aussie eclipse happens? Better start planning your trip!

THIS WAY UP
WARNING
DO NOT OPEN SPACE WINDOW!

7
Wave to the astronauts

Right now, a team of astronauts and cosmonauts* is whizzing above our heads at 27,580 km per hour. They are zooming along 420 km above Earth, safely sealed inside a spacecraft the size of a rugby pitch – an incredible Earth-orbiting laboratory where they live and work for up to 11 months at a time. It is called the International Space Station (ISS) and it has been permanently inhabited since the year 2000.

Working in space

The people working on the ISS are highly trained pilots, engineers and scientists. They conduct experiments to understand the effects of spaceflight on the human body and on cells, plants and insects. They study

* Several different words are used to describe a person whose full-time job is to train for spaceflights and work in space. These include 'cosmonaut' (Russia), 'taikonaut' (China), 'vyomanaut' (India) and 'spationaut' (France). Another term, 'spaceflight participant', describes someone who has flown in space but is not a full-time professional astronaut. This might include a space tourist or someone travelling on a short-term mission.

Earth's environment; grow fresh food like zucchinis, lettuce, radishes and wheat; and test how new materials perform when they are 'weightless', to help prepare for future missions to the Moon and Mars.

The sensation of weightlessness is also known as 'microgravity'. While standing on Earth, we feel heavy because the ground is pushing back at our bodies against the force of gravity. While orbiting Earth, we feel weightless because the floor of the spacecraft is also tumbling through space and can't push back at our bodies. It's like when you go down in a lift and your body feels lighter and your stomach goes all funny. Microgravity means that people working in space for long periods quickly lose the strength in their muscles. For that reason, they have to work out in a specially designed space-gym for 2 hours every day.

You can check out ISS video tours, space station experiments and live rocket launches on YouTube. Many of the astronauts share on social media the incredible pictures of Earth they take from space. You can also see the real thing with your own eyes. Orbiting Earth 16 times a day, the ISS is often visible in our skies. Go to the official webpage spotthestation.nasa.gov and type in the name of your own or nearest town or city. The website will tell you the dates and times of the next ISS flyover at your location. You can even set up an email or SMS alert to remind you when to look up, or track the current position of the ISS on your night sky app.

The ISS looks best when it's high overhead, so search for passes that have a maximum height of at least 40° – otherwise it will skim the horizon and get lost amid trees, buildings or light pollution. You will see what looks like a bright star moving slowly and steadily across the sky. And you won't need binoculars or a telescope – it's very clear with the naked eye. Don't forget to wave to the astronauts!

Identifiable flying objects

The ISS is not the only spacecraft that we can see in the night sky. Thousands of satellites orbit Earth, and many of them are visible in our skies as sunlight is reflected off their shiny surfaces and gleaming solar panels. There are also many discarded rocket casings that wander on their own above Earth, no longer used after their frantic work in propelling a satellite into orbit above our planet.

Most satellites are in a 'low-Earth orbit', meaning they are between 160 and 1000 km above Earth. Since satellites emit no light of their own, we can only see them when sunlight is reflected off their surfaces (and particularly their shiny solar panels), so the best time to see them is within an hour or two of sunset – during the night, these satellites go dark as they are plunged into Earth's shadow. So pick a time just after sunset, grab your night sky app, and find your favourite dark place to watch the skies for moving objects!

Satellites look like aeroplanes, except they are faster and have no flashing lights. Once you spot a satellite, you can point the night sky app at it and identify exactly what you're looking at, whether it's a communications satellite, a space telescope or a discarded rocket. You can even photograph satellites by taking a long-exposure snap – their long streaky light trails will stand out against the background of stars.

Satellite 'mega-constellations' such as the Starlink satellites move in large groups. This team of more than 1200 small spacecraft orbit Earth and beam fast wi-fi to homes and businesses around the world. They also freak a lot of people out – some even mistake the Starlink satellites for unidentified flying objects (UFOs) or aliens! If you want to spot a weird-looking bunch of lights trooping across the evening sky, you can check out findstarlink.com to calculate the next local pass of these alien sky-dancers!

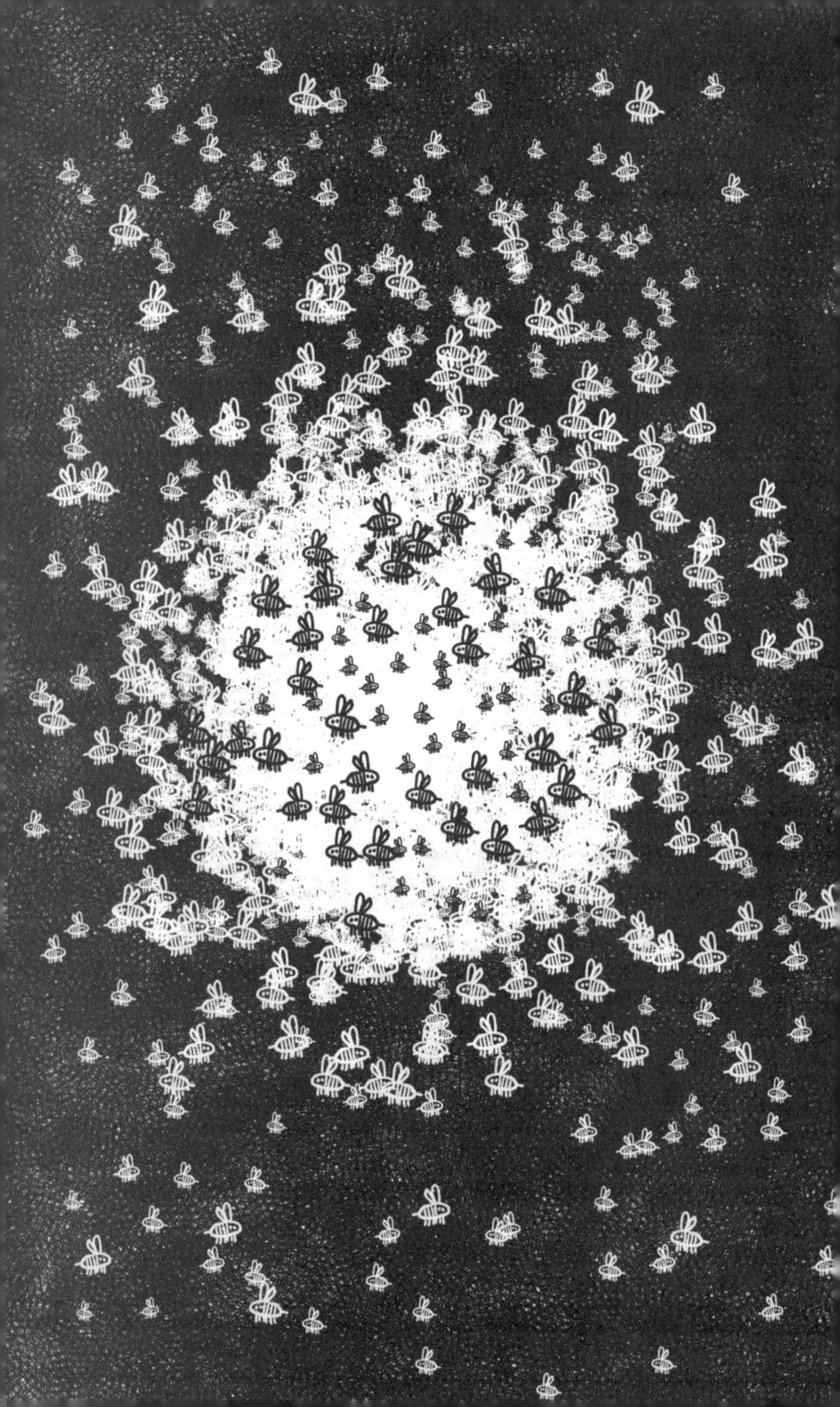

8

Travel back in time

Ever fancied yourself as a real-life time traveller? You can soar back in time simply by looking up at the night sky. Follow a straight line from Sirius through the brilliant white star Canopus to find 2 fuzzy smudges of light floating among the stars. These smudges are called the Large Magellanic Cloud (LMC) and the Small Magellanic Cloud (SMC).

Described in various Indigenous stories as an old man and woman sitting by a campfire, or as 2 wetland birds (cranes), the Magellanic Clouds are 2 small irregular galaxies[🚀] made up of hundreds to thousands of millions of stars.

The most amazing thing about the night sky is the distances involved. The LMC is 160,000 light-years[🪐] from Earth and the SMC is 210,000 light-years away. The light from this galaxy that you are seeing today

---

🚀 Galaxies like the Milky Way are called 'spiral' galaxies due to their shape. Other galaxies are described as 'elliptical' (oval) or 'irregular' (randomly shaped).

🪐 One light-year is 9.5 trillion km, or the distance that a beam of light travels in a year.

has been zooming through the universe for more than 200,000 years – it began its journey when Neanderthals were still walking the planet! Looking at the Magellanic Clouds is like travelling through time.

Scan the southern night sky with a pair of binoculars or a telescope and you will find hundreds of smudges of light. These 'deep sky objects' are often blurry and ill-defined because they represent some of the most distant items visible in our night sky. Use the search tool on your night sky app to help you find them.

The brightest and the best

Nebula (*neb*-you-la) is a Latin word meaning 'cloud'. Nebulae (*neb*-you-lee, meaning more than 1 nebula) are gigantic glowing clouds in space, often made up of a mixture of gases that spilled out of a star that died millions of years ago. As gravity pulls the gas together, new stars are born in a nebula.

The Orion Nebula is one of the easiest deep sky objects to observe with the naked eye since it is relatively close to Earth (just over 1300 light-years away) and contains around 700 bright young stars. To find it, look above Orion's Belt and note the 3 stars in a vertical line. This is called Orion's Sword because from the Northern Hemisphere (where it is the other way up) it seems to hang from his belt. Look carefully at the middle 'star' of

Orion's Sword and you will see that it looks like a fuzzy smudge of light – this is the Orion Nebula.

Using binoculars or a telescope, you will see that the Orion Nebula is cloud-like and contains a bright cluster of young blue stars called the Trapezium. To the naked eye, the nebula appears greyish-white, but a long-exposure photograph will reveal an amazing pink colour and dark patches of dust.

Many other nebulae are visible to us, including the Carina Nebula, the black Coalsack Nebula in Crux, the Tarantula Nebula in the LMC, and the Lagoon and Trifid nebulae in Sagittarius.

When groups of baby stars have finished forming in a nebula, they end up as large clusters of grown-up stars that blow away any surrounding gas and begin to shine freely. There are dozens of star clusters like this that are visible in our southern skies. The best way to view them is through binoculars or a telescope – it is a spectacular sight.

In summer you can enjoy the Pleiades and Hyades clusters. Follow Orion's Belt down to the bright orange star Aldebaran in the constellation of Taurus. The Hyades star cluster surrounds Aldebaran in a V-formation like the horns of a bull. Follow the line down from Orion's Belt and you will reach the Pleiades, also known as the Seven Sisters. The Pleiades has 6 or 7 bright blue stars visible to the naked eye (depending on your eyesight), but with binoculars you can see dozens more gleaming like jewels.

A long list of impressive young star clusters can be seen with binoculars, including the Southern Pleiades in Carina, the impressive pair of M46 and M47 in Puppis, and the Jewel Box cluster in Crux. Type their names into your night sky app to find them.

Swarming space-bees!

For the incredible sight of 10 million stars bound by gravity, you can't go past Omega Centauri, the brightest globular cluster in our skies. Find an imaginary point in the sky right between Crux and the Pointers (imagine the cross upright), then draw a straight line twice the length of the Southern Cross upwards from this point. You will see a smudge of light, which through binoculars or a telescope will immediately reveal itself to be a tight ball of millions of stars. It's a swarm of space-bees!

For extra homework, find the second-brightest globular cluster, called 47 Tucanae, which lies right next to the SMC.

A globular cluster is a very densely packed collection of extremely old stars that live far outside the disc of our galaxy.

For one last time travel adventure, why not try to glimpse the distant galaxy Centaurus A? At more than 10 million light-years away, it's the most distant object you're likely to see through a small pair of binoculars ... AND there's a supermassive black hole at its centre.

Once again, find the imaginary place in the sky that lies between Crux and the Pointers, with the Southern Cross facing upright. Now follow a straight line upwards approximately 3 times the height of the Southern Cross. Centaurus A looks like an elongated smudge of light through binoculars – it will have a dark patch of dust across the middle if you're using a telescope. This one can be a bit tricky, so make sure the sky is dark and use your night sky app to help locate it.

Go forth with your binoculars and night sky app, and explore!

9
Things that go bump in the night

Not everything in our night sky is slow and steady. Sometimes, objects that are not stars or planets come by to surprise and delight us. A comet unfurls its remarkable tail like a peacock displaying his feathers. Great streaks of starlight fall like giant raindrops from the sky. The aurora australis dances silently across the horizon. Let's learn how to observe them.

Catch a comet by the tail

Comets are collections of rock and ice[*] that range from a few hundred metres to the size of Sydney or Melbourne. They circle the Sun in highly eccentric orbits. That doesn't mean they wear a bowler hat and carry a pet salamander on their shoulder. No, I mean that unlike the near-circular paths taken by planets, the oval orbits of comets take them far from the Sun over hundreds or thousands of years before bringing them back for brief visits to the inner solar system.

---

[*] That's why comets are often called 'dirty snowballs'.

When a comet gets close to the Sun, its surface is warmed by the heat of our star. The ices heat up and become gases, which vent off into space to create the comet's fuzzy head (also called a coma) and tail. The comet becomes bright and can often be spotted from Earth for several weeks or even months before disappearing back into the cold darkness of space.

Whichever direction the comet is travelling in, the tail is pushed by a constant stream of high-energy particles escaping from the Sun, which we call the solar wind. That's why a comet's gas tail always points directly away from the Sun. On some comets, small particles and grains of rock lag behind the gas tail, forming a second 'dust tail'.

When I was a teenager, I was treated to not 1 but 2 great comets that were visible to the naked eye. In 1996, Comet Hyakutake appeared. It was brighter than the stars in Orion and had a long tail and a greenish colour. My first glimpse of this breathtaking sight over the rooftop of my childhood home was unforgettable. As if that wasn't enough, the following year, the greatest comet in a generation dropped by. Comet Hale-Bopp outshone Sirius and was clear to the naked eye for more than a year from my observation site in the small village of Wethersfield in south-eastern England. Its 2 tails were quite a sight!

When will the next naked-eye comet come to visit Earth? This is very difficult to predict, since comets' orbits typically take thousands of years to complete and they are only visible when they get close to the Sun. But I can tell you that Comet Leonard is predicted to reach naked-eye brightness between December 2021 and January 2022, and new comets are being discovered all the time. The best thing to do is keep an eye on the monthly online sky guides for the Southern Hemisphere.

A fantastic by-product of a comet is the trail of space dust it leaves behind. As Earth steadily orbits the Sun, our planet intersects several of these comet trails throughout the year. When Earth hits a comet trail, we experience a 'meteor shower' as thousands of shooting stars rain down into our skies. They are caused by particles of dust (most the size of a grain of sand) burning up in the atmosphere 50 to 100 km above the ground as they rub against the air at 50 km per second.

Glorious light shows

With a little patience and luck, you can see shooting stars any night of the year. Some create coloured trails of light, including yellows, blues, greens and reds, depending on the chemicals making up the meteor. If you really want to see a show, remember that some meteor showers appear like clockwork and peak at the same time every year. Most showers last a week or so, and the number of visible shooting stars will be much greater at these key times.

With its peak on or around 14 December, the Geminids meteor shower usually offers the best chance of seeing shooting stars in Australia. Looking up at dark skies in a part of the world where the source of the meteors (called the radiant), the constellation of Gemini, is directly above your head, you might expect to see as many as 150 meteors per hour. In Australia, though, the radiant is always fairly low in the sky, so many meteors will be below our horizon. The expected numbers in our southern skies are as many as 1 meteor every minute at the peak.

The northern part of Australia is the best place to see the Geminids. The radiant rises at around 9.30 pm in Darwin and 9 pm in Brisbane. In Perth, it rises at 10 pm. If you're in Sydney, Canberra, Melbourne, Adelaide or Hobart, you'll have to wait another hour or more before the show starts. Remember to let your eyes adapt to the dark, get comfy on a deckchair, and take it all in. Don't be tempted to set up binoculars or a telescope – they won't help you. Just use your eyes, as they have a much wider field of vision. If you want to try photographing meteors, set your camera to manual and open the shutter for 10–30 minutes. You'll almost certainly get an image with at least 1 meteor streaking across it.

Other large meteor showers to look out for are the Eta Aquarids (5–6 May) and the Orionids (21 October), and there are dozens of minor showers throughout the year. That's why it's always worth going somewhere dark and looking up! Your night sky app will give you the lowdown on exactly when and where to look from your location.

Finally, there is the rare and fabulous sight of the aurora australis. This is mainly one for you if you live in Tassie, but it is occasionally seen in the southern parts of Western Australia, South Australia, the Australian Capital Territory, New South Wales and Victoria.

The aurora australis, or the 'southern lights', is a phenomenal light show that happens high up in our atmosphere when high-energy particles from the Sun crash into molecules of nitrogen and oxygen and shake light out of them. These conditions often happen during times of solar activity, when the Sun is firing hot gas out into space. The aurora is seen above the north and south poles because that's where our planet's magnetic field channels these solar particles.

There are no guaranteed times or dates for viewing the southern lights. You just have to fluke it and be in a dark, southerly location while the Sun is cooking up a storm. Keep an eye on the forecast at the Bureau of Meteorology website (sws.bom.gov.au/Aurora), and if your location is under an aurora alert, look south towards the horizon and hope for the best – a south-facing beach is an ideal spot.

The naked eye sees a bright aurora as a shimmering grey cloud, sometimes with moving layers that appear to be dancing. You can also see vertical beams of light that look like searchlights. A fainter aurora might only be visible with a camera. Using manual mode (or your NightCap Camera app on a phone or tablet), you can capture spectacular green and/or red colours, even when nothing is visible to the naked eye. Good luck!

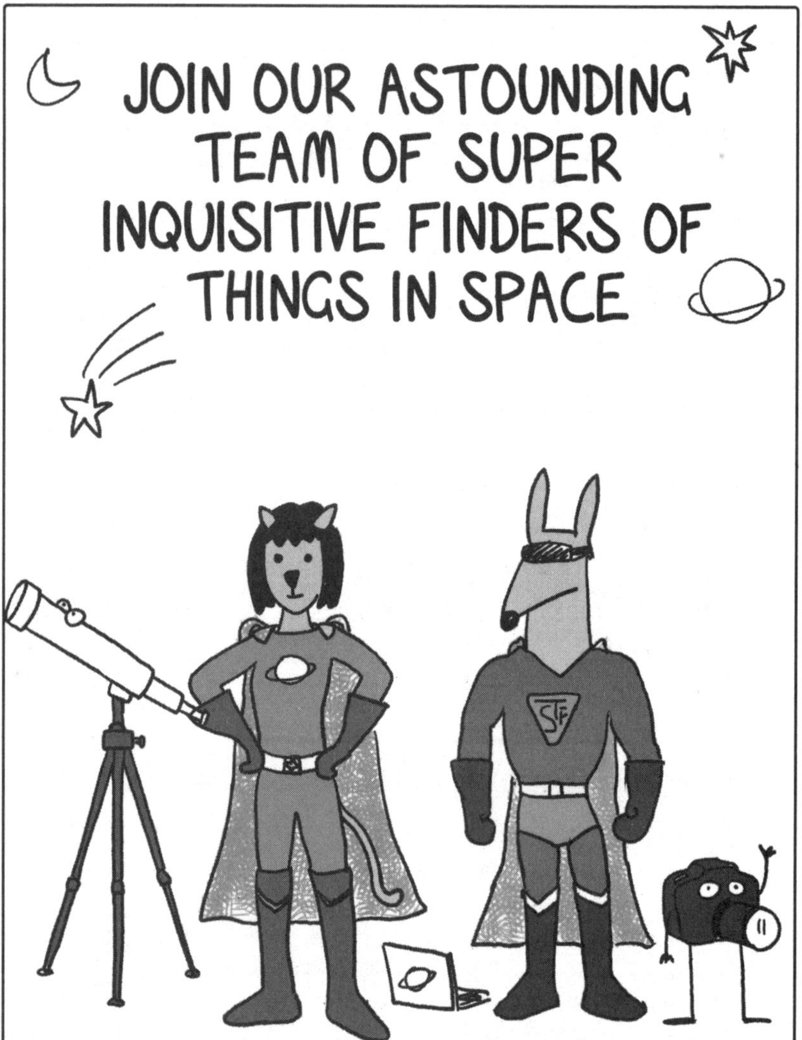
JOIN OUR ASTOUNDING TEAM OF SUPER INQUISITIVE FINDERS OF THINGS IN SPACE
THE SPACE THING FINDERS
Come and join us!

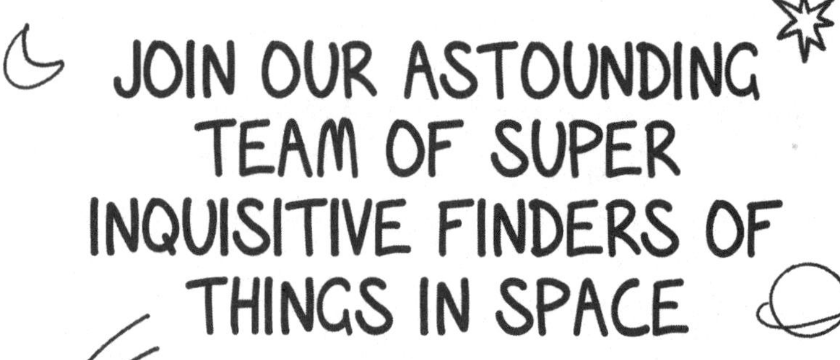

10
Supersize your hobby

If you have enjoyed exploring the night sky and now want to supersize your hobby, I strongly recommend that you visit your local amateur astronomical society.[*] Apart from organising talks and having members with a wealth of expertise on hand, they often run stargazing nights with telescopes available for use. I joined my local group (the Braintree Astronomical Society) in England when I was 12 and I never looked back. There, I met passionate amateur astronomers, listened to fascinating talks by professional astronomers, and enjoyed field trips to the London Planetarium. I also learned about astrophotography and telescopes, visiting a local observatory and borrowing another member's telescope for a few weeks. Years later, after studying at university, I became a professional astronomer myself!

Many beginners want to buy a telescope but don't know where to start. Here are some basic rules to guide you.

---

[*] You can find a list of contacts here: astronomy.org.au/amateur/amateur-societies/australia.

First, don't rush out and buy a cheap 'toy' telescope. A good pair of binoculars and a sturdy tripod will serve you far better. But if you have already learned your way around the night sky with binoculars and now want to go deeper into the universe, then by all means save up for a good telescope that will bring spiral galaxies, faint star clusters and planetary moons into your sights.

Telescopes are named according to their optics (that's the lenses, mirrors and eyepiece) and the type of mount (the legs or base that hold up the telescope and allow it to point in different directions). What is the best telescope for you? There isn't one right answer. It depends on how much money you have to spend, how much storage space you have, and whether you want to mainly look at objects or photograph them. Try different telescopes at your local astronomy club and talk to the members who already own one – they can give you pointers about what's right for you. Remember that telescopes are cheaper second-hand, so shop for a used one if you can.

My advice is to save up for a Dobsonian telescope with an aperture of 150 mm or greater. A Dobsonian is a reflecting telescope, meaning it has a large mirror at the bottom that reflects light from the stars onto a smaller mirror inside the tube and then out through an eyepiece containing a lens, which magnifies the light and focuses it into your eye. The telescope tube is attached to an alt-az mount,[*] which spins around on its base and pivots up and down so you can easily point the telescope at anywhere in the sky.

For a cheaper version, you might consider a 76 mm (or preferably larger) tabletop reflecting telescope with a Dobsonian mount. You can purchase a brand-new one online for around $140. They are very portable, meaning you can store one in your bedroom and take it outside at night. Just be aware that at this price, the image brightness and quality may be limited. Best to try before you buy.

If your budget is a bit bigger (or you are willing to save up for longer), you can get a brand new 150 mm aperture Dobsonian telescope for around $500, or a second-hand version for less. This will bring planets to life, transport you to the Moon's craters, and enable you to see many deep-sky objects like the Orion Nebula, Omega Centauri and other star clusters. Just make sure that you have somewhere dry and safe to keep it, like a garden shed. And ensure it comes with finder scope – this is a small telescope attached to the tube of your big

[*] When looking at a telescope mount, 'alt-az' means altitude (up and down) and azimuth (side-to-side) motion. This distinguishes it from an equatorial mount, which swings around the South Celestial Pole.

scope that allows you to easily find and point yourself at an object before you look through the main eyepiece.

The magnification (how much bigger the object looks) and field of view (how much of the sky you can see) will depend on the eyepiece you are using. Telescope eyepieces can easily be swapped, so it is best to have a few different ones for looking at different objects. The quality of the eyepiece is important, so if your telescope has a decent-sized mirror but the images are no good, then upgrading the eyepiece is a sensible place to start. Ask at your local astronomy society or in a specialist telescope or camera shop for advice.

If night sky photography is your passion, you also have the option of buying a camera mount for your telescope. Designs vary, but basically they hold the camera still at the eyepiece so you can take close-up pictures of stars, moons, planets and galaxies. If you're using a phone camera, buy a phone mount. If you have a big, chunky digital SLR camera, you will need a DSLR mount.

An interest in astronomy should last a lifetime – there is always something new to discover. Binoculars, telescopes and cameras are great for maximising your hobby, but you can explore a meteor shower, a comet, the aurora australis or a new spacecraft passing overhead just by using your eyes. The tools and technologies will change too – and that's all part of the fun.

Research some constellations that are visible tonight. Don't be afraid to try different tools and techniques – you will be amazed at how quickly you improve. Find as many planets as you can. Set the alert to wave at the International Space Station. Watch a meteor shower. Gaze back in time at the Magellanic Clouds. Sigh at the beauty of the Jewel Box star cluster. Watch the next full Moon rise spectacularly in the east against city lights, an ocean swell or moonlit hills.

Enjoy our night sky. Share it and protect it for future generations.

And never stop exploring.

YOU LOOKING
AT SATURN

SATURN
LOOKING
AT YOU!

Fun facts

New stars that are being born often shoot laser beams of microwaves into space. Yep, you heard me right. These beams are called masers, and they help us study what's going on inside the clouds that are birthing young stars.

*

The Andromeda galaxy is the closest spiral galaxy to the Milky Way. Not only that, but it's heading towards us at 400,000 km per hour. In around 3.8 billion years, the galaxies will collide!

*

Since Mars and Earth are both orbiting the Sun, the distance between the planets changes constantly. Their separation can be as little as 54.6 million km at closest approach, and up to 401 million km apart at their furthest point. That's more than 7 times further apart!

*

Earth's seasons happen because Earth is tilted at 23.5° from upright relative to the orbit of our planet around the Sun. It probably wasn't always that way. When the solar system was formed, our planet almost certainly had no orbital tilt. It was probably caused when Earth collided with a huge protoplanet (a young planet that is still forming) billions of years ago.

Night sky apps are great, but looking at the bright screen of your phone will ruin your night vision. Most apps (including Sky Guide) have a night mode, which turns the screen dark and the words and images red. This will help to maintain your eyes' superpowers while you navigate the sky. Try it!

✷

The first telescope was designed by a Dutch spectacle maker named Hans Lippershey in 1608. The story goes that some children were playing in his workshop and found that if they lined up 2 lenses in the right way, they could see a faraway weathervane get bigger. He took the idea and made lots of money from the invention. Tsk – grown-ups, eh?

✷

In ancient China, many people believed that a solar eclipse represented a dragon eating the Sun. During an eclipse, they would bang drums to scare the dragon away. It always worked!

✷

The diameter of a telescope is important because you want to let more light inside so you can see faraway, faint objects. The world's largest optical telescope is located on the island of La Palma in the Canary Islands and has a diameter of 10.4 m.

Stars make up only 4% of the matter (stuff) in the Milky Way. Another 12% of the galaxy is made up of gas. The remaining 84% of the Milky Way is 'dark matter', a mysterious substance we don't yet understand. Spooky or what?

*

If you live at the equator, with a clear horizon you can see every star in our night sky – whether it's in the Northern or Southern Hemisphere. From the South Pole, you'll only ever see the southern sky, and from the North Pole, it's the northern sky only. The rest of us in between can see parts of the northern and southern skies throughout the year.

*

On Mercury, temperatures soar to more than 400 °C during the day. At night, they plummet to an icy -180 °C. Why so extreme? The lack of an atmosphere on Mercury means that heat is not trapped close to the planet, so temperatures vary a lot.

*

Venus spins in the opposite direction to Earth. We don't know why.

*

Jupiter has 79 moons. Probably the most exciting is Io (pronounced *eye*-oh), which is so volcanic that it regularly spews lava into space!

The moons of Mars, called Phobos and Deimos, are asteroids that were captured by Mars's gravity. Unlike regular moons, they are not round in shape. Instead, they look more like potatoes.

✷

Neptune was discovered after scientists found that the orbit of Uranus was deviating from its predicted path. They realised that an unseen planet must be 'pulling' Uranus off course with its gravity. It was the first planet to be discovered using mathematics.

✷

Aboriginal and Torres Strait Islander people were the world's first astronomers, developing ways to observe the Sun, Moon and stars. This knowledge informs navigation, calendars and weather prediction. First Nations cultures stretch back 65,000 years and continue today.

✷

The aurora has spectacular colours when photographed but usually looks white to the naked eye. Why? It's because the human eye's colour receptor cells, called cones, do not work well in darkness. A camera sensor is very sensitive to colour, and therefore the greens, reds and purples look stunning when photographed.

✷

Only 12 human beings have ever walked on the Moon – and all of these events took place between 1969 and 1972.

Glossary

Aperture
The size of the opening in a camera or telescope. The larger the aperture, the more light can enter the device, and therefore fainter objects can be seen.

Asteroid
A rocky body in the solar system that orbits the Sun.

Astronomer
A person who studies the universe to understand more about stars, planets, moons, asteroids, comets, galaxies, etc. Astronomers can be professional scientists (astrophysicists), or amateur astronomers who look at the stars for fun.

Astrophysicist
A professional scientist who studies the universe using maths and physics, and careful measurements, to make predictions and understand objects and forces in space.

Aurora
A beautiful light show that happens when gas belched out of the Sun interacts with Earth's atmosphere. It is frequently visible from polar regions (close to the North Pole or South Pole) during times when the Sun is active, but it's occasionally seen from other areas too.

Axis

An imaginary straight line that joins the north and south poles of a star, planet or moon.

Celestial

In the night sky, or in outer space.

Comet

A celestial object made of rock and frozen chemical 'ices' that travels around the Sun in a long and stretched-out orbit. When it comes near the Sun, the ices melt and turn into gases, often forming a tail that is visible from Earth.

Compass

A device that aligns with Earth's magnetic field to show the user which direction is north or south.

Constellation

A group of stars, which may be named after the imagined shape of an object, person, animal or mythical creature.

Cosmic

Outside of Earth, or in the cosmos (universe).

Diagonal

A line that cuts across the middle of a shape from one corner to the other.

Diameter

The width of a circle from one side to the other.

Eclipse

An astronomical event where one object (like the Sun or Moon) covers another, either directly with its body or with its shadow.

Equator

An imaginary circle around Earth that runs at an equal distance from the north and south poles.

Galaxy

A gigantic collection of stars that are bound together in space by gravity.

Gravity

The force that pulls objects towards each other. Gravity is what causes things to fall to the ground on Earth. It is also responsible for the orbits of moons and planets.

Hemisphere

Half of a sphere. The word usually refers to the 'top' or 'bottom' half of Earth (the Northern Hemisphere or Southern Hemisphere).

Horizon

The line where the land (or sea) meets the sky.

Meteor

A streak of light in the night sky caused by a small piece of space rock burning up in Earth's atmosphere.

Microgravity
The feeling of weightlessness experienced by astronauts in orbit around Earth.

Neanderthals
An extinct species of human that used basic tools and medicine. Scientific evidence suggests that they became extinct around 40,000 years ago.

Nebula
A cloud of gas in space that is visible from Earth due to the reflection of light from nearby stars, or by its own glowing light.

Orbit
The motion of objects around one another caused by the force of gravity.

Particle
A tiny building block of material that makes up everything.

Planet
A roughly spherical (basketball-shaped) body made of rock, liquid or gas that orbits a star.

Planisphere
A map of the stars printed on a circular disc that allows you to see which stars are visible at any particular time or date.

Rocket casing

The large metal tube that holds rocket fuel when a spacecraft is launched into space. Once the fuel has all been burned, the metal tube sometimes remains in orbit and can be seen from Earth as a fast-moving 'star', because it reflects sunlight.

Satellite

An object that orbits a larger body in space. Artificial satellites are spacecraft made by humans that are launched into space and circle Earth. Moons are natural satellites.

Star

A spherical collection of gas in space that releases heat and light by the process of nuclear fusion, which combines particles to give off energy.

Universe

Everything that exists.

Resources

Here's a handy list of the websites and apps in the book.

Help fight light pollution in Australia:
australasiandarkskyalliance.org

Pick a perfect stargazing location near you with this dark-places map:
darksitefinder.com

Find out the current lunar phase:
timeanddate.com/moon/phases

See an up-to-date map of the night sky from the Sydney Observatory or Perth's Scitech museum (choose the one closest to you):
maas.museum/observations/category/monthly-sky-guides

scitech.org.au/explore/the-sky-tonight

Stargaze and planet hop like a pro with these night sky apps:
Sky Guide ($4.49, available for iPhone)
SkyView Lite (free, but has fewer features; available for iPhone and Android)

Capture images of stars, planets, star trails and even the southern lights with this photo app:
NightCap Camera ($4.49, available for iPhone)

Look up the position of planets in your area tonight:
timeanddate.com/astronomy/night

Explore the surface of the Moon:
quickmap.lroc.asu.edu

Note: the default map shows a view of the Moon in the Northern Hemisphere – select the Lunar Globe (3D) projection and drag the image to match the Moon as it appears in the Australian sky. Select the 'nomenclature' overlay to see the names of maria and craters.

Find out when the International Space Station will be flying over your location next:
spotthestation.nasa.gov

Look up the location of the Starlink satellites, a mega-constellation of spacecraft that beam fast wi-fi to Earth:
findstarlink.com

Keep an eye on the aurora forecast for your area at the Bureau of Meteorology website:
sws.bom.gov.au/Aurora

Join your local amateur astronomical society:
astronomy.org.au/amateur/amateur-societies/australia

Acknowledgements

In the pursuit of this wonderful hobby (and later career) of astronomy, I have benefited from the wisdom and kindness of many people. My dad first showed me the stars as a young child, and my mum drove me across the wilds of Essex, England, to meetings of the Braintree Astronomical Society and observatory nights in various cold, dark, windy locations. Members of the Braintree Astronomical Society showed patience and care in nurturing my interest and helping me learn how to use telescopes and cameras to capture the cosmos, as did Linda Simonian of the Astronomy Centre in Todmorden. Talented astronomy authors, including Patrick Moore, Heather Couper and Nigel Henbest, fed my fascination through their insatiable writing energy. To all of them, I owe my love of the stars.

This book was made possible by the work of a number of talented publishers, editors and artists, including Sally Heath, Paul Smitz, Jessica Levine, Eugenie Baulch, Phil Campbell and, of course, the inimitable First Dog on the Moon. It was a pleasure to work with you all.

Enjoy the journey.

Professor Lisa Harvey-Smith is an award-winning astronomer, Australia's Ambassador for Women in Science, Technology, Engineering and Mathematics (STEM), and a professor at the University of New South Wales. She has a master's degree in astronomy and astrophysics and a PhD in radio astronomy. Lisa is the author of *When Galaxies Collide* (2018), *Under the Stars* (2019), *The Secret Life of Stars* (2020) and *Aliens and other Worlds: True tales from our solar system and beyond* (October 2021). She is a regular on our screens and was a presenter on ABC TV's *Stargazing Live.* She has toured Australia with several live shows – including 3 featuring Apollo-missions astronauts Buzz Aldrin, Charlie Duke and Gene Cernan – and with her own stage show, *When Galaxies Collide*. In her spare time, she loves running up mountains in the mud. For more on Lisa and her work, visit lisaharveysmith.com.

READING MAKES YOU HEALTHIER, SMARTER AND HAPPIER.*

READ MORE BOOKS, MORE OFTEN.

'I virtually lived in Braintree library (in Essex, England) when I was a kid. My family would spend hours browsing and reading before borrowing our maximum allowance of 6 books for the week. I read almost every book in the place before I was 18! However busy you are, I reckon there is always time before bed to read a few pages. In my experience a few pages turn into a few chapters and I end up nodding off and dropping my iPad on my face.'

LISA HARVEY-SMITH

This book is proudly published by Thames & Hudson Australia to celebrate Australian Reading Hour.

Australian Reading Hour is your official excuse to stop whatever you're doing (school, homework, chores), pick up a book – and read for an hour!

Get involved, join the fun, and choose more books to read at **australiareads.org.au**

Australia Reads is on a mission to get more people reading more books more often. We are a not-for-profit reading initiative working in collaboration with members of the Australian Booksellers Association, Australian Library and Information Association, Australian Publishers Association, and Australian Society of Authors to champion reading.

***seriously. Make reading for fun a healthy daily habit, like exercising for 30 mins, or drinking plenty of water, or cleaning your teeth twice each day. Put down your phone, pick up a book. Turn off the TV and turn over this page...**

---

australiareads.org.au

Venture further into the cosmos with these other books by Professor Lisa Harvey-Smith

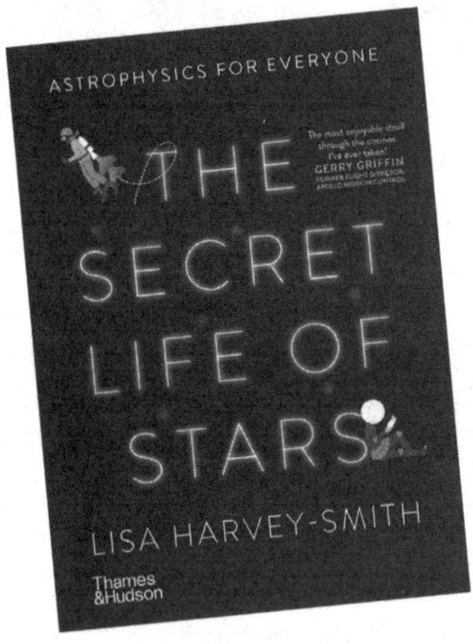

Out October 2021

Visit us at: thamesandhudson.com.au

Follow us on Instagram: @thamesandhudsonau
and Facebook: @thameshudsonaustralia